Even or Odd?

A number is **even** if you can show it as pairs with nothing left over.

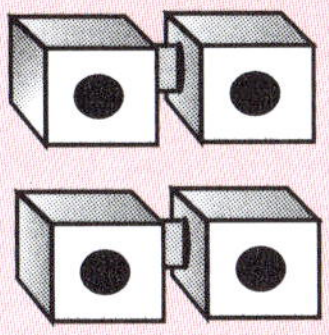

4

So, 4 is an even number.

A number is **odd** if you can show it as pairs with one left over.

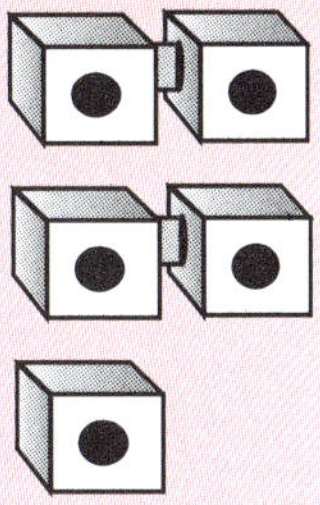

5

So, 5 is an odd number.

Choose the even number.

1. 3 or 6
2. 12 or 19
3. 35 or 36
4. 7 or 10
5. 28 or 15
6. 21 or 4

Choose the odd number.

7. 15 or 10
8. 4 or 7
9. 21 or 28
10. 19 or 36
11. 12 or 3
12. 35 or 6

Answer Box

A	B	C	D	E	F
15	19	36	3	35	7
G	**H**	**I**	**J**	**K**	**L**
10	6	28	21	4	12

Objective: Identify a number as odd or even.

Model Building

Find the number of hundreds, tens, or ones.

1. 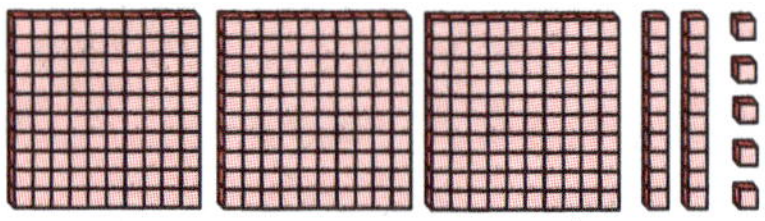
3 hundreds + ■ + 5 ones

2. 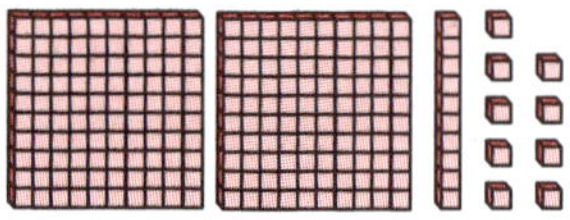
2 hundreds + 1 ten + ■

3. 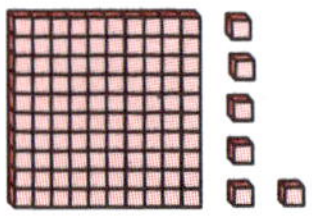
1 hundred + ■ + 6 ones

4. 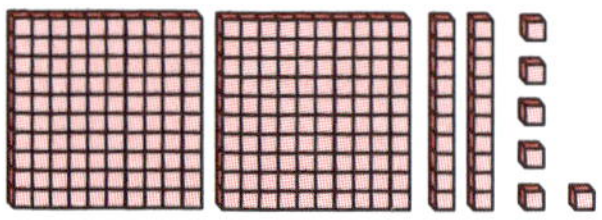
2 hundreds + 2 tens + ■

5. 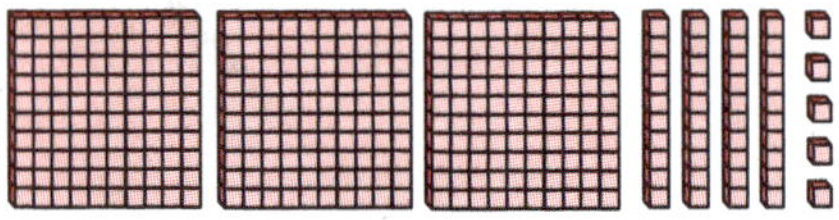
3 hundreds + ■ + 5 ones

6. 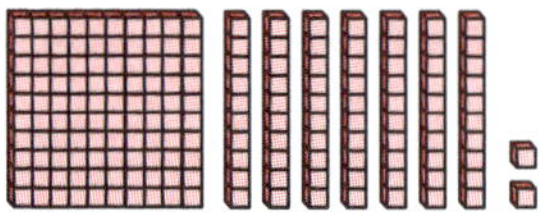
■ + 7 tens + 2 ones

7. 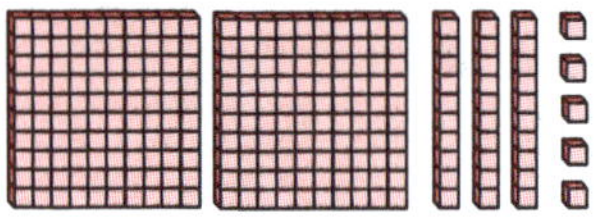
■ + 3 tens + 5 ones

8. 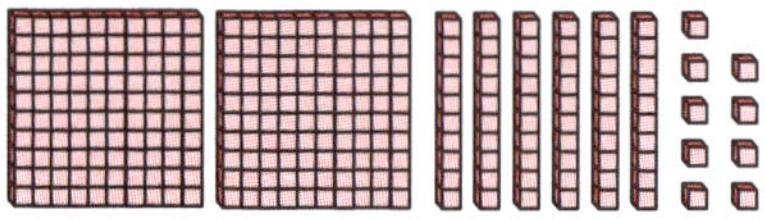
2 hundreds + ■ + 9 ones

9. 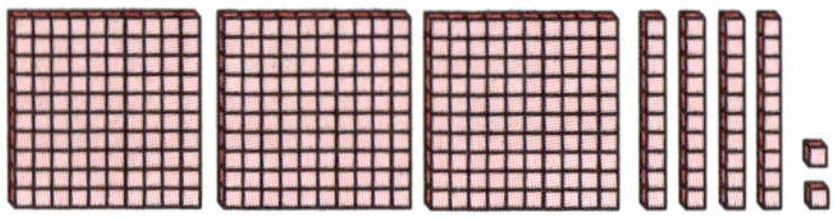
■ + 4 tens + 2 ones

10. 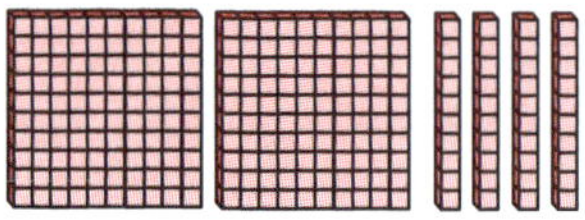
2 hundreds + 4 tens + ■

11. 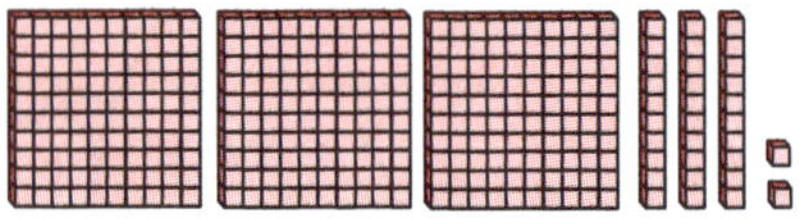
3 hundreds + ■ + 2 ones

12. 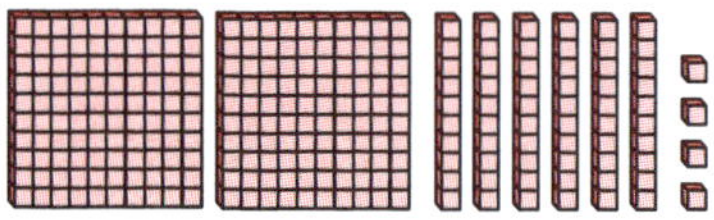
2 hundreds + 6 tens + ■

Answer Box

A	B	C	D	E	F
2 hundreds	0 ones	4 ones	6 tens	4 tens	3 tens
G	**H**	**I**	**J**	**K**	**L**
2 tens	6 ones	9 ones	1 hundred	3 hundreds	0 tens

Objective: Name the number of hundreds, tens, and ones in a group of place-value models.

What's My Name?

Find the number.

1 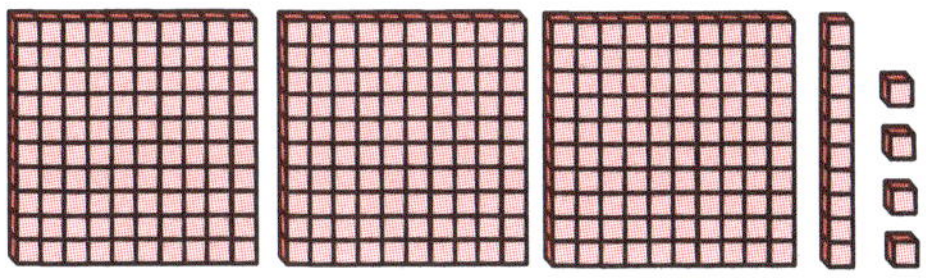

2 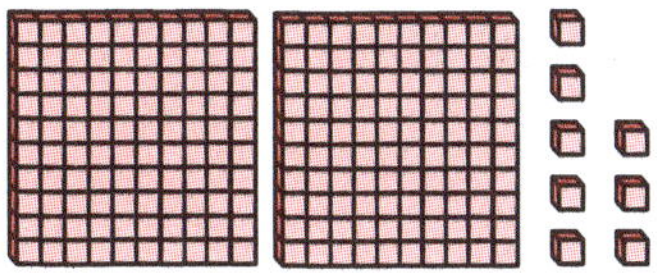

3 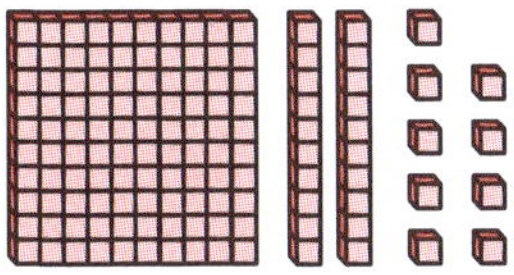

4 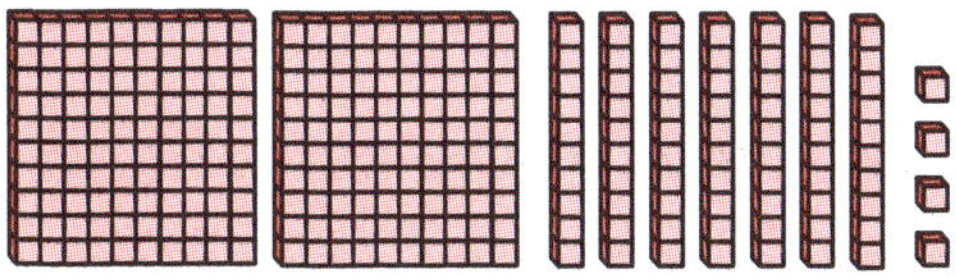

5 

6 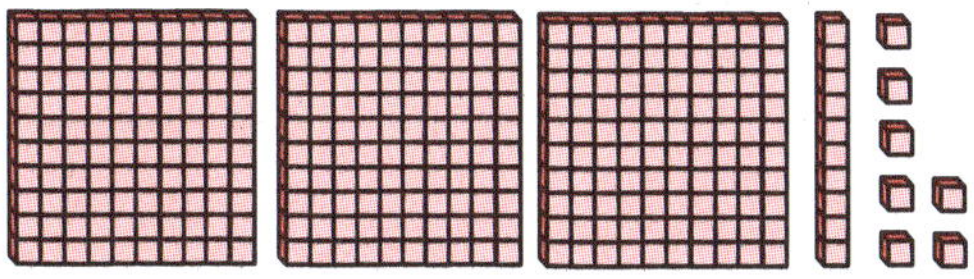

7 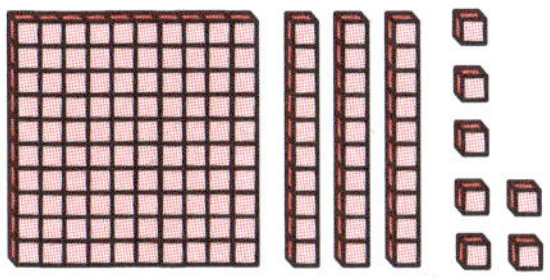

8 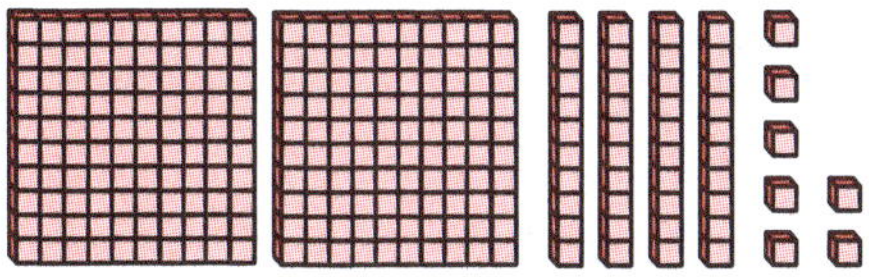

9 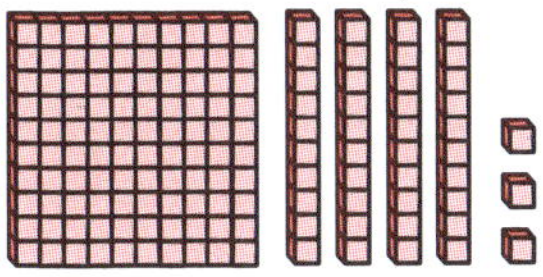

10 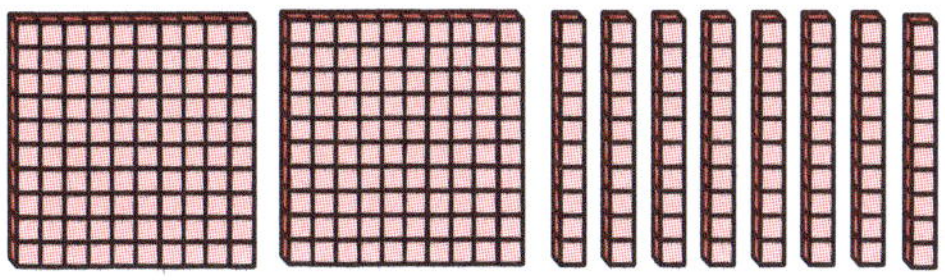

11 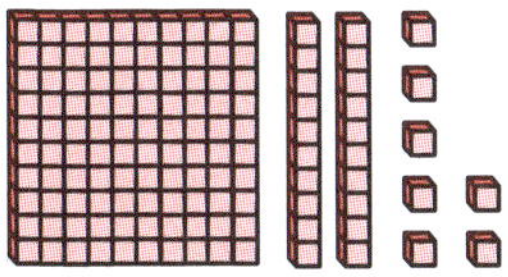

12 

Answer Box

A	B	C	D	E	F
129	127	274	280	317	247
G	**H**	**I**	**J**	**K**	**L**
193	143	137	314	183	208

Objective: Name a 3-digit number, given a place-value model.

Valuable Numbers

Example

Find the value of the digit that is underlined and in color.

Hundreds	Tens	Ones
4	2	3

So, 2 tens = **20**.

Find the value of the underlined digit.

	Hundreds	Tens	Ones
1	4	3	1
2	5	6	8
3	7	0	5
4	1	8	6
5	7	5	3
6	2	9	4
7	8	6	3
8	4	3	9
9	3	8	6
10	3	8	7
11	9	6	0
12	8	7	4

Answer Box

A	B	C	D	E	F
3	900	70	9	800	7
G	**H**	**I**	**J**	**K**	**L**
700	300	90	8	80	30

Objective: Name the value of an underlined digit in a 3-digit number shown in a place-value chart.

Bigger and Bigger

Hundreds	Tens	Ones
2	4	7

Standard form: 247   Expanded form: 200 + 40 + 7

Find the missing number.

	Hundreds	Tens	Ones		
1	5	2	7	→	500 + ■ + 7
2	7	4	1	→	■ + 40 + 1
3	1	8	2	→	100 + 80 + ■
4	2	6	5	→	200 + ■ + 5
5	4	9	6	→	■ + 90 + 6
6	8	1	9	→	800 + ■ + 9
7	3	5	7	→	300 + 50 + ■
8	6	4	8	→	■ + 40 + 8
9	2	0	7	→	200 + ■ + 7
10	6	3	9	→	600 + 30 + ■
11	4	9	8	→	400 + ■ + 8
12	7	4	6	→	700 + ■ + 6

Answer Box

A	B	C	D	E	F
600	40	0	2	9	7
G	**H**	**I**	**J**	**K**	**L**
10	700	90	400	60	20

Objective: Complete the expanded form for a 3-digit number shown in a place-value chart.

What's My Value?

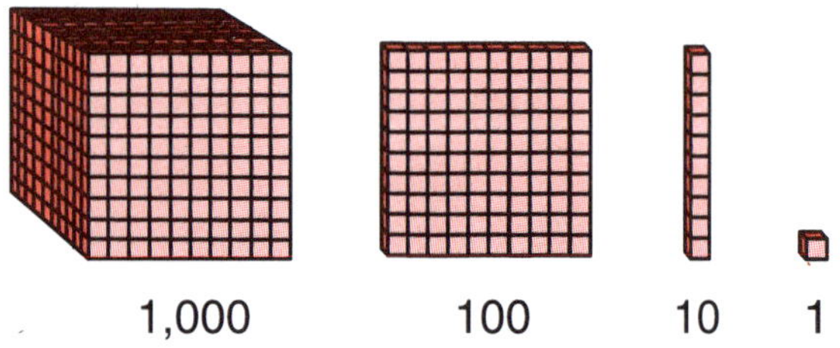

Find the value of the digit that is underlined and in color.

	Thousands	Hundreds	Tens	Ones
1	6	4	8	3
2	4	1	2	9
3	1	3	8	6
4	8	5	1	4
5	9	0	7	1
6	2	4	2	0
7	5	4	9	2
8	8	1	0	5
9	2	8	6	4
10	8	5	3	7
11	7	7	4	6
12	3	9	7	0

Answer Box

A	B	C	D	E	F
400	9	5,000	8,000	1,000	4
G	**H**	**I**	**J**	**K**	**L**
900	100	80	500	70	700

Objective: Name the value of an underlined digit in a 4-digit number shown in a place-value chart.

Watch Your Place!

Find the missing number.

	Thousands	Hundreds	Tens	Ones	
1	3	4	5	8	→ 3,000 + 400 + ■ + 8
2	5	8	2	6	→ 5,000 + ■ + 20 + 6
3	2	9	2	0	→ 2,000 + 900 + ■ + 0
4	4	1	8	6	→ ■ + 100 + 80 + 6
5	9	6	2	5	→ 9,000 + 600 + 20 + ■
6	7	2	6	8	→ 7,000 + 200 + ■ + 8
7	8	4	1	8	→ 8,000 + ■ + 10 + 8
8	6	4	9	2	→ ■ + 400 + 90 + 2
9	3	5	7	1	→ 3,000 + ■ + 70 + 1
10	5	8	2	4	→ 5,000 + 800 + 20 + ■
11	4	8	1	2	→ 4,000 + 800 + 10 + ■
12	8	9	4	7	→ ■ + 900 + 40 + 7

Answer Box

A	B	C	D	E	F
50	4,000	500	20	800	60
G	**H**	**I**	**J**	**K**	**L**
4	400	5	2	6,000	8,000

Objective: Complete the expanded form for a 4-digit number shown in a place-value chart.

Valuable Places

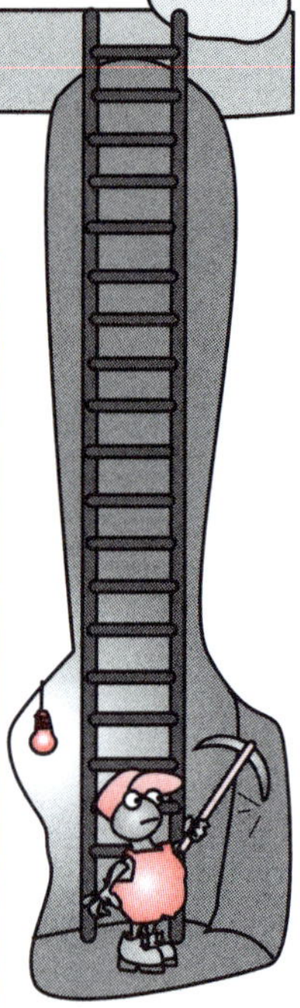

Find the value of the digit that is underlined and in color.

	Hundred Thousands	Ten Thousands	Thousands	Hundreds	Tens	Ones
1	4	5	**6**	5	2	1
2	5	2	4	**8**	0	3
3	6	**4**	7	1	3	7
4	2	8	5	0	9	**1**
5	**9**	7	3	7	5	6
6	8	3	2	**9**	4	3
7	4	7	6	0	**1**	7
8	1	2	0	**6**	8	4
9	7	**9**	4	3	2	8
10	3	**8**	1	2	7	0
11	2	4	**9**	7	3	2
12	6	5	**4**	5	9	8

Answer Box

A	B	C	D	E	F
10	6,000	4,000	800	900,000	90,000
G	**H**	**I**	**J**	**K**	**L**
1	80,000	900	600	40,000	9,000

Objective: Name the value of an underlined digit in a 5- or 6-digit number shown in a place-value chart.

Before, After, or Between?

Find the number that comes just before.

1. 225	2. 600	3. 6,421
4. 5,108	5. 21,001	6. 425,835

Find the number that comes just after.

7. 425,834	8. 5,108	9. 21,999
10. 589	11. 6,420	12. 225

Answer Box

A	B	C	D	E	F
6,420	22,000	5,107	224	425,834	599
G	**H**	**I**	**J**	**K**	**L**
6,421	21,000	590	425,835	5,109	226

Find the number that comes between.

1. 34, ■, 36	2. 60, ■, 62	3. 98, ■, 100
4. 60,599, ■, 60,601	5. 1,854, ■, 1,856	6. 20,642, ■, 20,644
7. 89, ■, 91	8. 600,597, ■, 600,599	9. 1,845, ■, 1,847
10. 62, ■, 64	11. 20,633, ■, 20,635	12. 349, ■, 351

Answer Box

A	B	C	D	E	F
99	1,846	20,643	600,598	60,600	35
G	**H**	**I**	**J**	**K**	**L**
20,634	1,855	350	61	63	90

Objective: Name the number that comes before, after, or between two given numbers through hundred thousands.

Line Up!

Find the missing number.

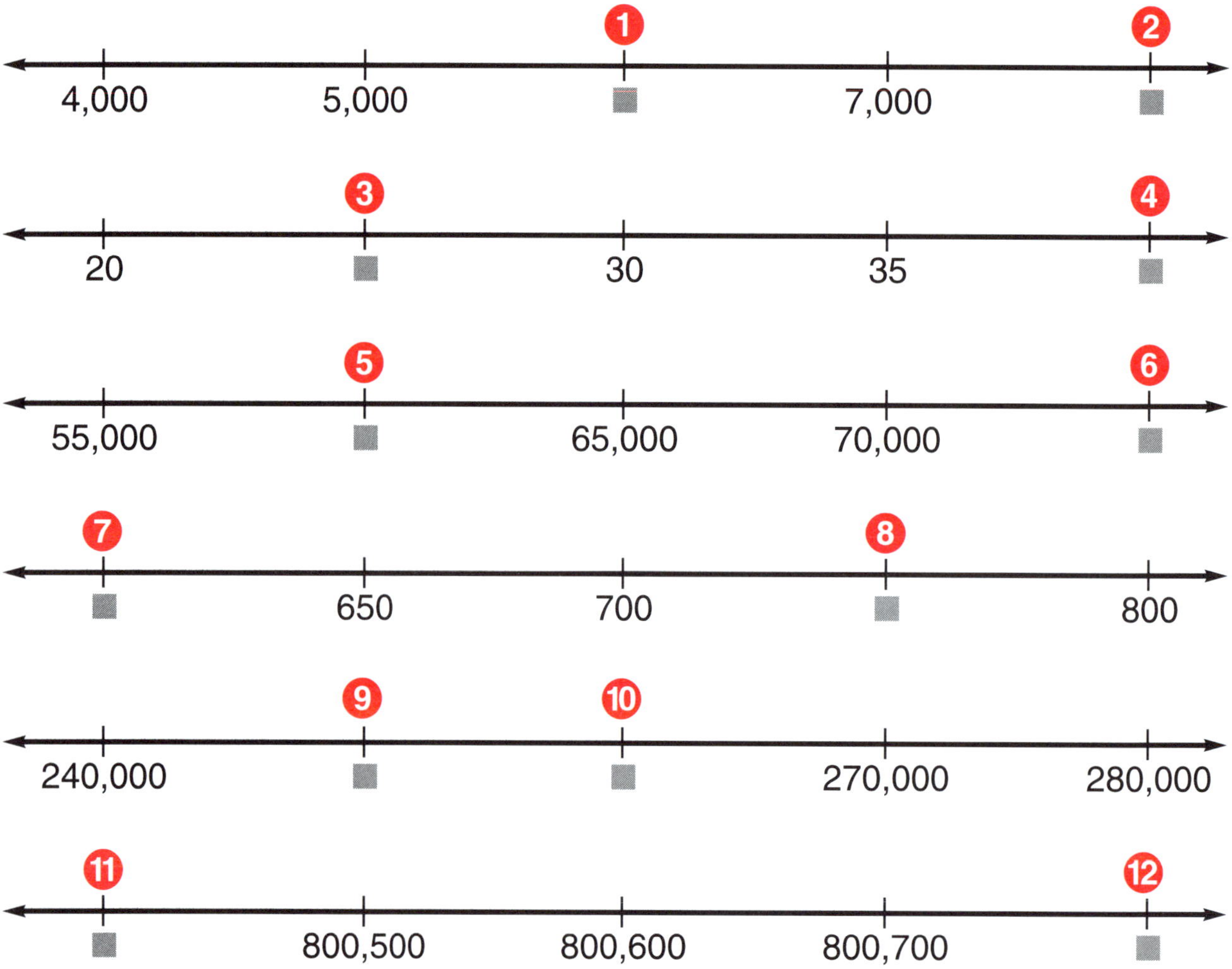

Answer Box

A	B	C	D	E	F
8,000	75,000	250,000	25	6,000	40
G	**H**	**I**	**J**	**K**	**L**
750	800,800	800,400	60,000	260,000	600

Objective: Use a number line to order numbers through hundred thousands.

Compare . . . If You Dare!

Example

Choose the number that makes the statement true.	Line up the numbers:	Compare the digits:
345 > ■	345	4 > 3
354 or 330	354	So, 345 > **330**.
	330	

Choose the number that makes the statement true.

1. 58 > ■
   47 or 74
2. 260 > ■
   265 or 256
3. 852 < ■
   940 or 825
4. 47 < ■
   74 or 47
5. 625 > ■
   2,065 or 265
6. 2,085 > ■
   2,580 or 825
7. 999 < ■
   2,058 or 256
8. 2,580 = ■
   2,580 or 2,058
9. 1,002 < ■
   940 or 2,065
10. 23,523 > ■
    23,532 or 23,325
11. 23,532 = ■
    233,532 or 23,532
12. 233,235 < ■
    23,325 or 233,532

Answer Box

A	B	C	D	E	F
233,532	2,580	2,058	47	265	23,532
G	**H**	**I**	**J**	**K**	**L**
256	825	74	23,325	2,065	940

Objective: Use >, <, or = to compare numbers through hundred thousands.

Problem Solving: Using a Table

Use the information in the table to answer the question.

The Summer Olympic Games were held in Atlanta, Georgia, from July 19 through August 4, 1996. The table gives the medal count for the top 10 medal winners in the 1996 Summer Olympics.

1996 OLYMPIC MEDALS

Country	Gold	Silver	Bronze	Total
United States	44	32	25	101
Germany	20	18	27	65
Russia	26	21	16	63
China	16	22	12	50
Australia	9	9	23	41
France	15	7	15	37
Italy	13	10	12	35
South Korea	7	15	5	27
Cuba	9	8	8	25
Ukraine	9	2	12	23

[Source: 1998 World Almanac]

1. Which country won the most medals in the 1996 Olympics?
2. Three countries won the same amount of bronze medals. How many medals is that?
3. Which country won the fewest silver medals?

4. Which country won the most bronze medals?
5. How many countries won a total of more than 40 medals?
6. How many countries won a total of less than 35 medals?
7. Cuba won 8 gold medals. True or false?
8. France won more gold medals than Italy. True or false?
9. You can use this table to find out how many gold medals Norway won. Yes or no?
10. You can use this table to find out the total amount of medals China won. Yes or no?
11. Which country won between 12 and 14 gold medals?
12. How many countries won more than 15 silver medals?

Answer Box

A	B	C	D	E	F
Yes	False	Italy	5	True	4
G	**H**	**I**	**J**	**K**	**L**
Germany	United States	Ukraine	No	3	12

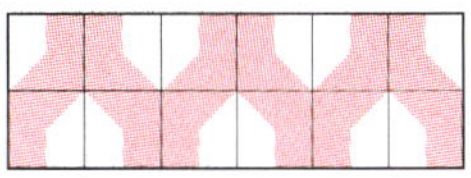

Objective: Use information from a table to solve a problem that involves ordering numbers.

One Hundred . . . More or Less!

Example

Find the number that is 100 more.

427
↑ Think: To increase by 100, add 1 to the digit in the hundreds place.

So, **5**27 is 100 more than **4**27.

Find the number that is 100 less.

427
↑ Think: To decrease by 100, subtract 1 from the digit in the hundreds place.

So, **3**27 is 100 less than **4**27.

Find the number that is 100 more.

1. 500
2. 640
3. 1,200
4. 8,547
5. 24,692
6. 534,400

Find the number that is 100 less.

7. 640
8. 610
9. 7,647
10. 2,300
11. 25,792
12. 535,500

Answer Box

A	B	C	D	E	F
600	8,647	534,500	740	7,547	1,300
G	**H**	**I**	**J**	**K**	**L**
540	2,200	535,400	510	24,792	25,692

Objective: Name the number that is 100 more than or 100 less than a given number.

One Thousand . . . More or Less!

Example

Find the number that is 1,000 more.

5,162

↑ Think: To increase by 1,000, add 1 to the digit in the thousands place.

So, **6**,162 is 1,000 more than **5**,162.

Find the number that is 1,000 less.

5,162

↑ Think: To decrease by 1,000, subtract 1 from the digit in the thousands place.

So, **4**,162 is 1,000 less than **5**,162.

Find the number that is 1,000 <u>more</u>.

1. 5,000
2. 3,279
3. 7,408
4. 29,400
5. 73,180
6. 141,700

Find the number that is 1,000 <u>less</u>.

7. 6,100
8. 5,379
9. 7,418
10. 84,180
11. 30,500
12. 152,700

Answer Box

A	B	C	D	E	F
29,500	8,408	6,000	5,100	4,379	142,700
G	**H**	**I**	**J**	**K**	**L**
6,418	74,180	83,180	30,400	4,279	151,700

Objective: Name the number that is 1,000 more than or 1,000 less than a given number.

Shady Parts

Example

Find the fraction that tells how much is shaded.

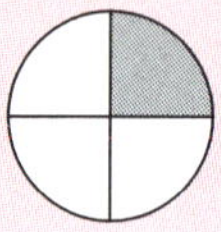

Think: 1 part is shaded.
There are 4 equal parts.

So, $\frac{1}{4}$ is shaded.

Find the fraction that tells how much is shaded.

1. 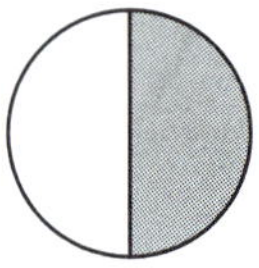
2. 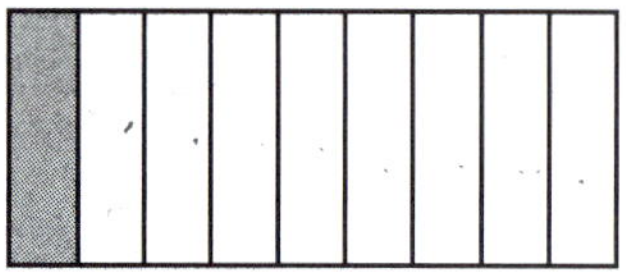
3. 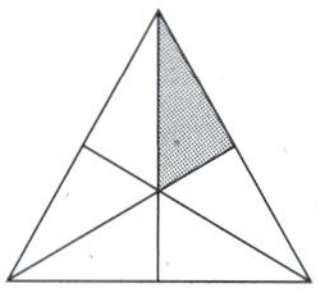
4. 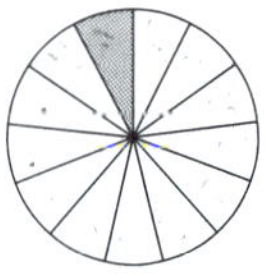
5. 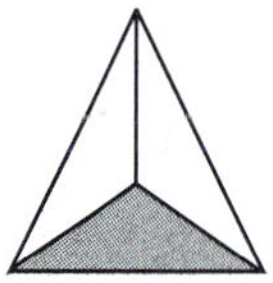
6. 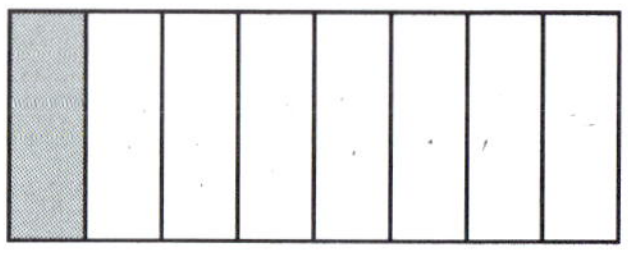
7. 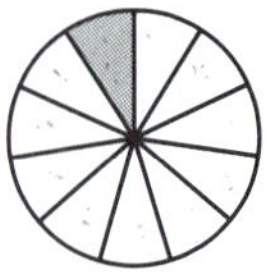
8. 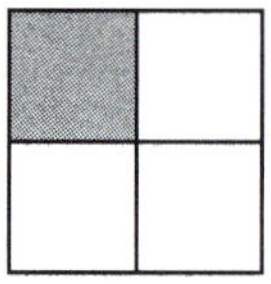
9. 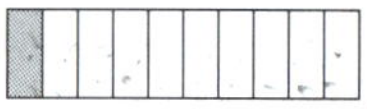
10. 
11. 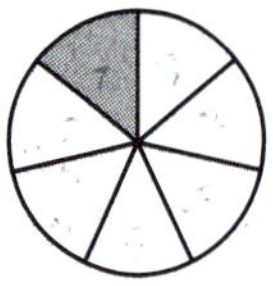
12. 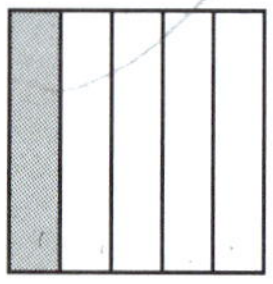

Answer Box

A	B	C	D	E	F
$\frac{1}{7}$	$\frac{1}{3}$	$\frac{1}{4}$	$\frac{1}{5}$	$\frac{1}{12}$	$\frac{1}{11}$
G	**H**	**I**	**J**	**K**	**L**
$\frac{1}{10}$	$\frac{1}{6}$	$\frac{1}{9}$	$\frac{1}{8}$	$\frac{1}{2}$	$\frac{1}{13}$

Objective: Name the unit fraction that tells how much of a region is shaded.

Be a Part of It All!

Example

Find the fraction that tells how much is shaded.

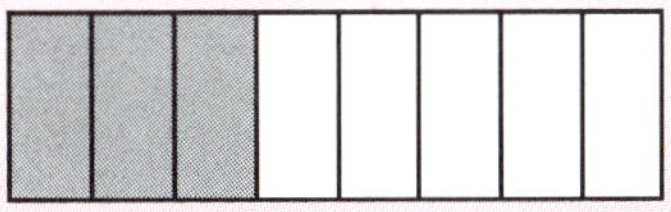

Think: 3 parts are shaded.
There are 8 equal parts.

So, $\frac{3}{8}$ is shaded.

Find the fraction that tells how much is shaded.

1 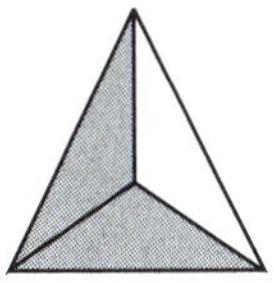

2 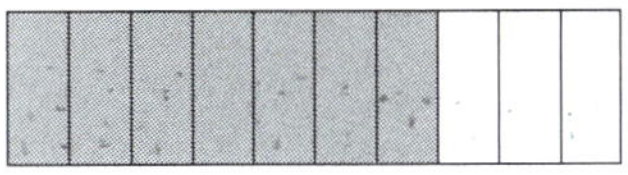

3 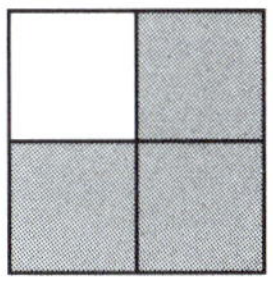

4 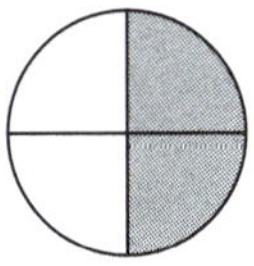

5 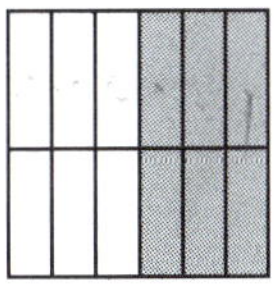

6 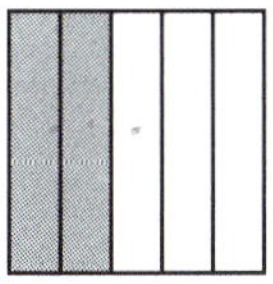

7 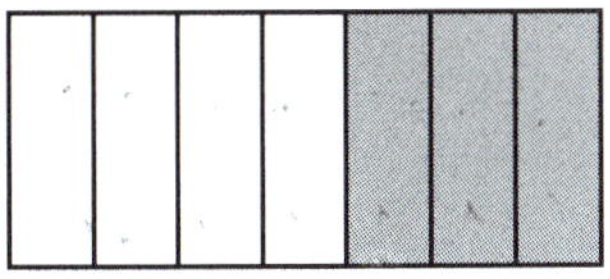

8 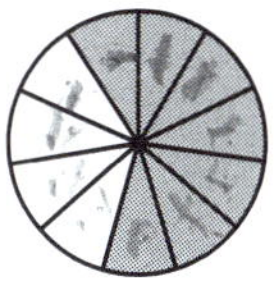

9 

10 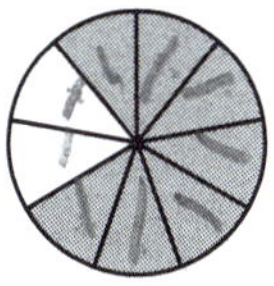

11 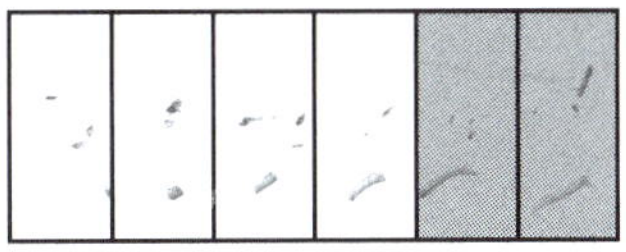

12 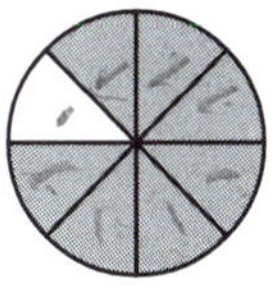

Answer Box

A	B	C	D	E	F
$\frac{2}{5}$	$\frac{7}{11}$	$\frac{3}{10}$	$\frac{3}{4}$	$\frac{3}{7}$	$\frac{2}{3}$
G	**H**	**I**	**J**	**K**	**L**
$\frac{7}{8}$	$\frac{7}{10}$	$\frac{6}{12}$	$\frac{2}{6}$	$\frac{2}{4}$	$\frac{7}{9}$

Objective: Name the non-unit fraction that tells how much of a region is shaded.

Part of the Group

Example

Find the fraction that tells what part of the group is shaded.

Think: 5 frogs are shaded.
There are 6 frogs.

So, $\frac{5}{6}$ is shaded.

Find the fraction that tells what part of the group is shaded.

1. 
2. 
3. 
4. 
5. 
6. 
7. 
8. 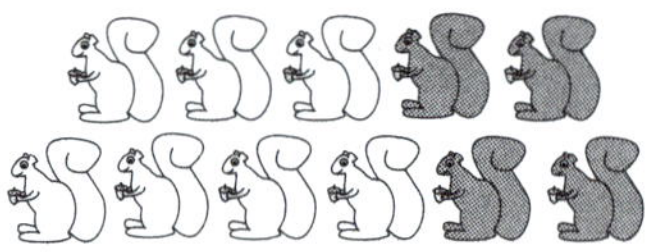
9. 
10. 
11. 
12. 

Answer Box

A	B	C	D	E	F
$\frac{4}{5}$	$\frac{1}{4}$	$\frac{5}{8}$	$\frac{2}{5}$	$\frac{4}{9}$	$\frac{3}{5}$
G	**H**	**I**	**J**	**K**	**L**
$\frac{3}{6}$	$\frac{4}{11}$	$\frac{3}{8}$	$\frac{5}{12}$	$\frac{3}{4}$	$\frac{1}{3}$

Objective: Name the fraction that tells how much of a group of objects is shaded.

Same Denominators ... Let's Compare!

Example

Choose the fraction that makes the statement true.	Look at the denominators.	Compare the numerators.
$\frac{3}{5} > ■$	$\frac{3}{5}, \frac{2}{5}, \frac{4}{5}$	$3 > 2$
$\frac{2}{5}$ or $\frac{4}{5}$	They are all the same.	So, $\frac{3}{5} > \frac{2}{5}$.

Choose the fraction that makes the statement true.

1. $\frac{3}{4} > ■$ — $\frac{2}{4}$ or $\frac{3}{4}$
2. $\frac{2}{6} < ■$ — $\frac{1}{6}$ or $\frac{5}{6}$
3. $\frac{7}{9} > ■$ — $\frac{5}{9}$ or $\frac{8}{9}$
4. $\frac{8}{10} < ■$ — $\frac{7}{10}$ or $\frac{9}{10}$
5. $\frac{2}{3} = ■$ — $\frac{1}{3}$ or $\frac{2}{3}$
6. $\frac{9}{12} > ■$ — $\frac{7}{12}$ or $\frac{11}{12}$
7. $\frac{1}{6} = ■$ — $\frac{5}{6}$ or $\frac{1}{6}$
8. $\frac{8}{10} > ■$ — $\frac{9}{10}$ or $\frac{7}{10}$
9. $\frac{10}{12} < ■$ — $\frac{11}{12}$ or $\frac{7}{12}$
10. $\frac{2}{4} < ■$ — $\frac{3}{4}$ or $\frac{2}{4}$
11. $\frac{2}{3} > ■$ — $\frac{1}{3}$ or $\frac{2}{3}$
12. $\frac{6}{9} < ■$ — $\frac{5}{9}$ or $\frac{8}{9}$

Answer Box

A	B	C	D	E	F
$\frac{1}{6}$	$\frac{2}{4}$	$\frac{2}{3}$	$\frac{1}{3}$	$\frac{7}{12}$	$\frac{7}{10}$
G	**H**	**I**	**J**	**K**	**L**
$\frac{9}{10}$	$\frac{3}{4}$	$\frac{5}{9}$	$\frac{11}{12}$	$\frac{8}{9}$	$\frac{5}{6}$

Objective: Use >, <, or = to compare fractions that have the same denominators.

Same Name

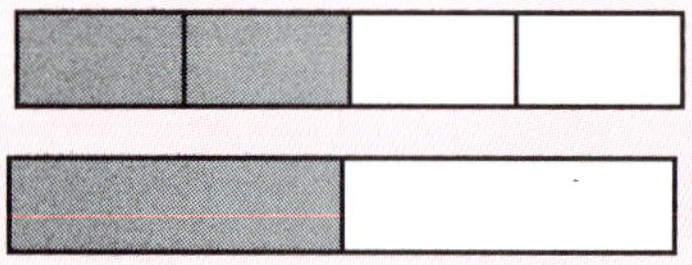

$\frac{2}{4}$

$\frac{1}{2}$

$\frac{2}{4}$ and $\frac{1}{2}$ are **equivalent fractions.**

They name the same amount.

Find the number that makes an equivalent fraction.

1. 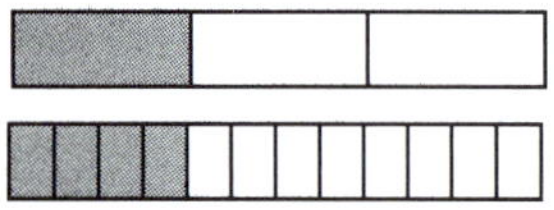 $\frac{1}{3} = \frac{\square}{12}$

2. 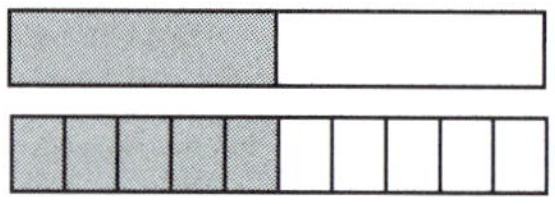 $\frac{1}{2} = \frac{\square}{10}$

3. 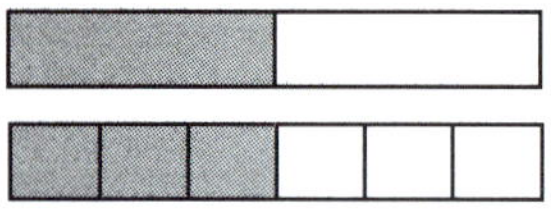 $\frac{1}{2} = \frac{\square}{6}$

4. 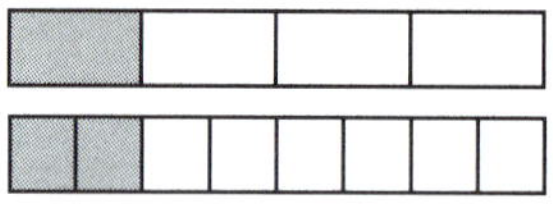 $\frac{1}{4} = \frac{2}{\square}$

5. 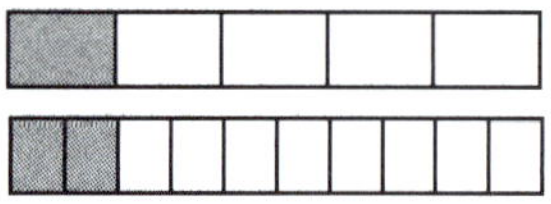 $\frac{1}{5} = \frac{2}{\square}$

6. 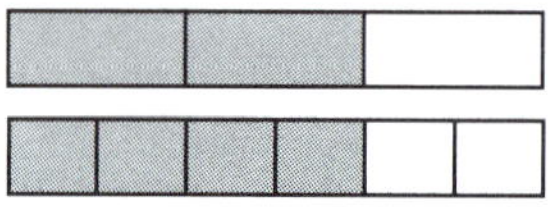 $\frac{2}{3} = \frac{4}{\square}$

7. 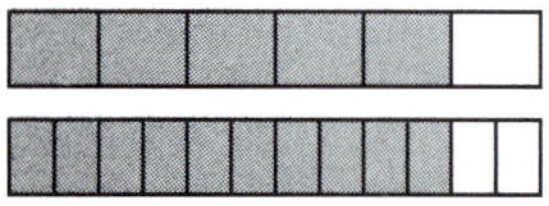 $\frac{5}{6} = \frac{10}{\square}$

8. 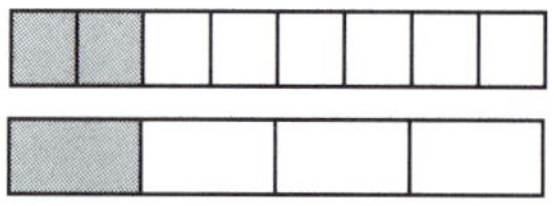 $\frac{2}{8} = \frac{\square}{4}$

9. 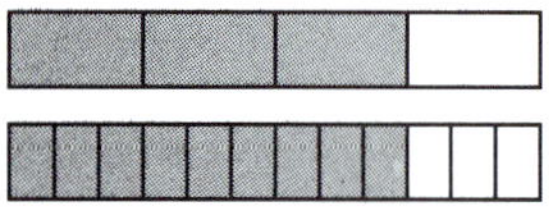 $\frac{3}{4} = \frac{\square}{12}$

10. 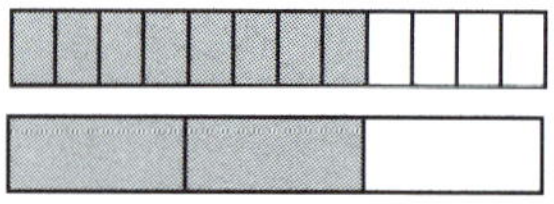 $\frac{8}{12} = \frac{\square}{3}$

11. 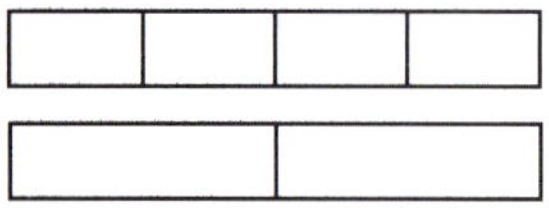 $\frac{0}{4} = \frac{\square}{2}$

12. 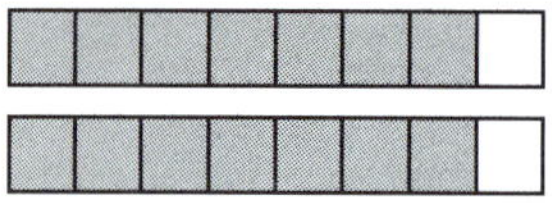 $\frac{7}{8} = \frac{\square}{8}$

Answer Box

A	B	C	D	E	F
2	8	7	1	9	10
G	**H**	**I**	**J**	**K**	**L**
4	12	5	6	0	3

 Objective: Complete an equivalent fraction.

Different Denominators . . . Let's Compare!

Choose the fraction that makes the statement true.

1 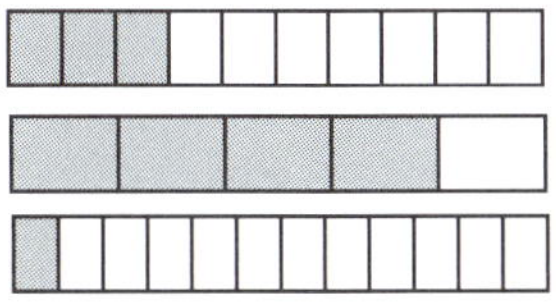

$\frac{3}{10} < \square$

$\frac{4}{5}$ or $\frac{1}{12}$

2 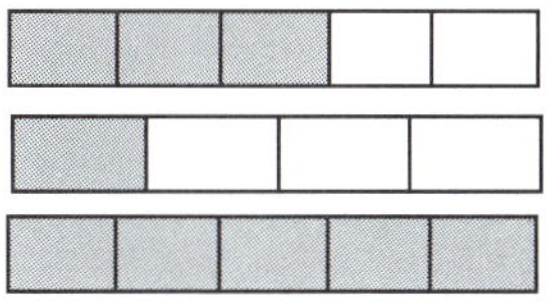

$\frac{3}{5} > \square$

$\frac{1}{4}$ or $\frac{5}{5}$

3 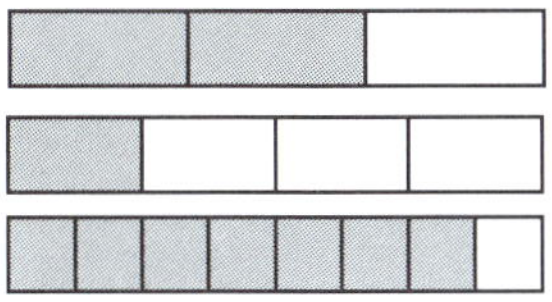

$\frac{2}{3} < \square$

$\frac{1}{4}$ or $\frac{7}{8}$

4 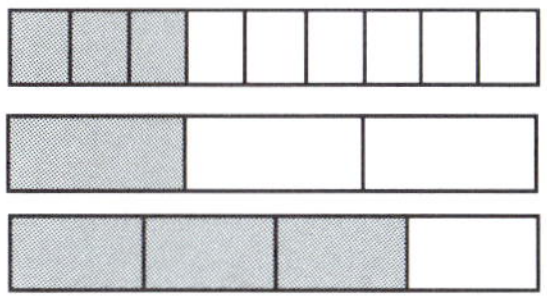

$\frac{3}{9} = \square$

$\frac{1}{3}$ or $\frac{3}{4}$

5 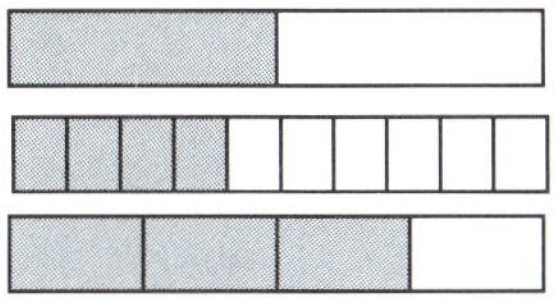

$\frac{1}{2} < \square$

$\frac{4}{10}$ or $\frac{3}{4}$

6 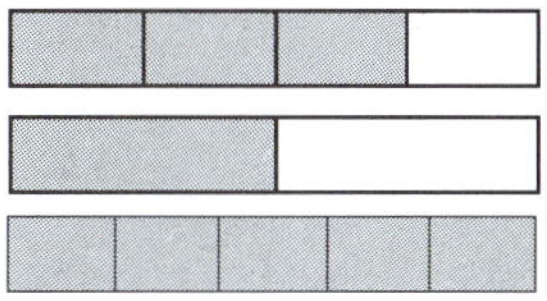

$\frac{3}{4} < \square$

$\frac{1}{2}$ or $\frac{5}{5}$

7 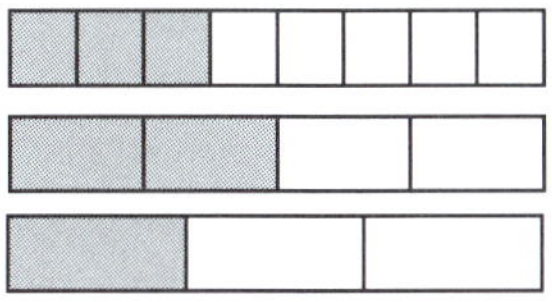

$\frac{3}{8} < \square$

$\frac{2}{4}$ or $\frac{1}{3}$

8 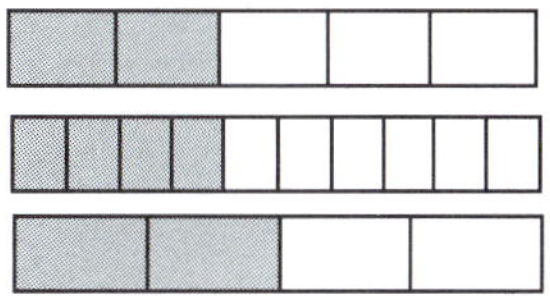

$\frac{2}{5} = \square$

$\frac{4}{10}$ or $\frac{2}{4}$

9 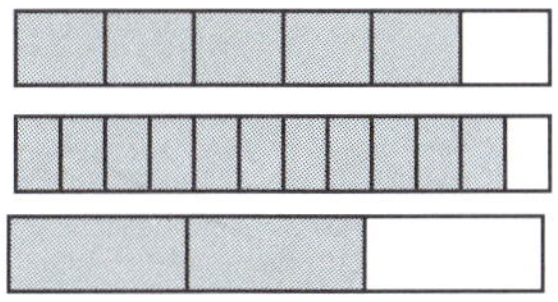

$\frac{5}{6} > \square$

$\frac{11}{12}$ or $\frac{2}{3}$

10 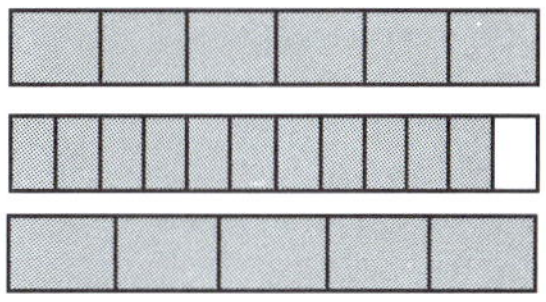

$\frac{6}{6} > \square$

$\frac{11}{12}$ or $\frac{5}{5}$

11 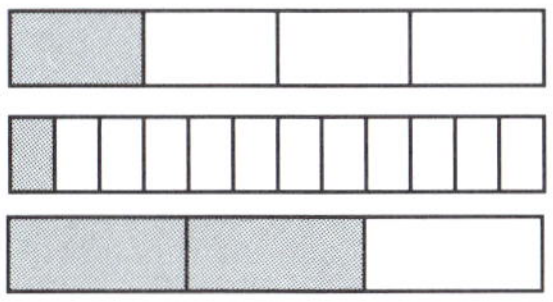

$\frac{1}{4} > \square$

$\frac{1}{12}$ or $\frac{2}{3}$

12 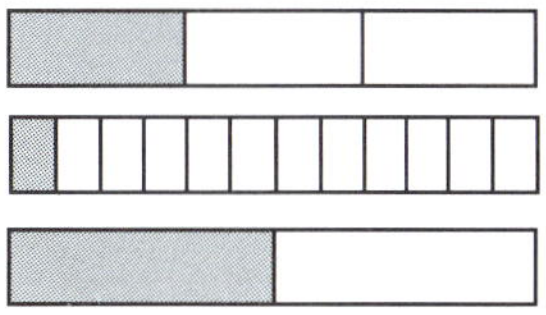

$\frac{1}{3} < \square$

$\frac{1}{12}$ or $\frac{1}{2}$

Answer Box

A	B	C	D	E	F
$\frac{2}{3}$	$\frac{3}{4}$	$\frac{11}{12}$	$\frac{2}{4}$	$\frac{4}{10}$	$\frac{5}{5}$
G	**H**	**I**	**J**	**K**	**L**
$\frac{1}{12}$	$\frac{7}{8}$	$\frac{1}{3}$	$\frac{4}{5}$	$\frac{1}{4}$	$\frac{1}{2}$

Objective: Use >, <, or = to compare fractions that have different denominators.

Problem Solving: Using Fractions

Use the picture to help solve the problem.

1. What fraction of the bikes are not on the track?
2. What fraction of the bikes have two wheels?
3. What fraction of the bikes that are on the track are in the front row?
4. What fraction of the bikes that are on the track are in the back row?
5. Of the bikes that are off the track, $\frac{2}{5}$ have a flat tire. True or false?
6. Of all the bikes, $\frac{2}{10}$ have a flat tire. True or false?
7. What fraction of the bikes that are off the track have handlebar streamers?
8. What fraction of the flag on the left has stripes?
9. What fraction of the flag on the right does not have stars?
10. Find the greater fraction: the part of the bikes that have streamers or the part of the bikes that do not have streamers.
11. Find the lesser fraction: the white part of the flag on the left or the white part of the flag on the right.
12. Find the lesser fraction: the part of the bikes on the track with flat tires or the part of the bikes off the track without flat tires.

Answer Box

A	B	C	D	E	F
$\frac{5}{12}$	$\frac{2}{7}$	$\frac{2}{4}$	$\frac{1}{7}$	$\frac{12}{12}$	False
G	**H**	**I**	**J**	**K**	**L**
$\frac{2}{5}$	$\frac{7}{12}$	True	$\frac{1}{3}$	$\frac{0}{7}$	$\frac{2}{3}$

Objective: Solve a problem that involves describing a part of a region or a group or that involves comparing two fractional amounts.

Fraction Line Up

Find the missing fraction.

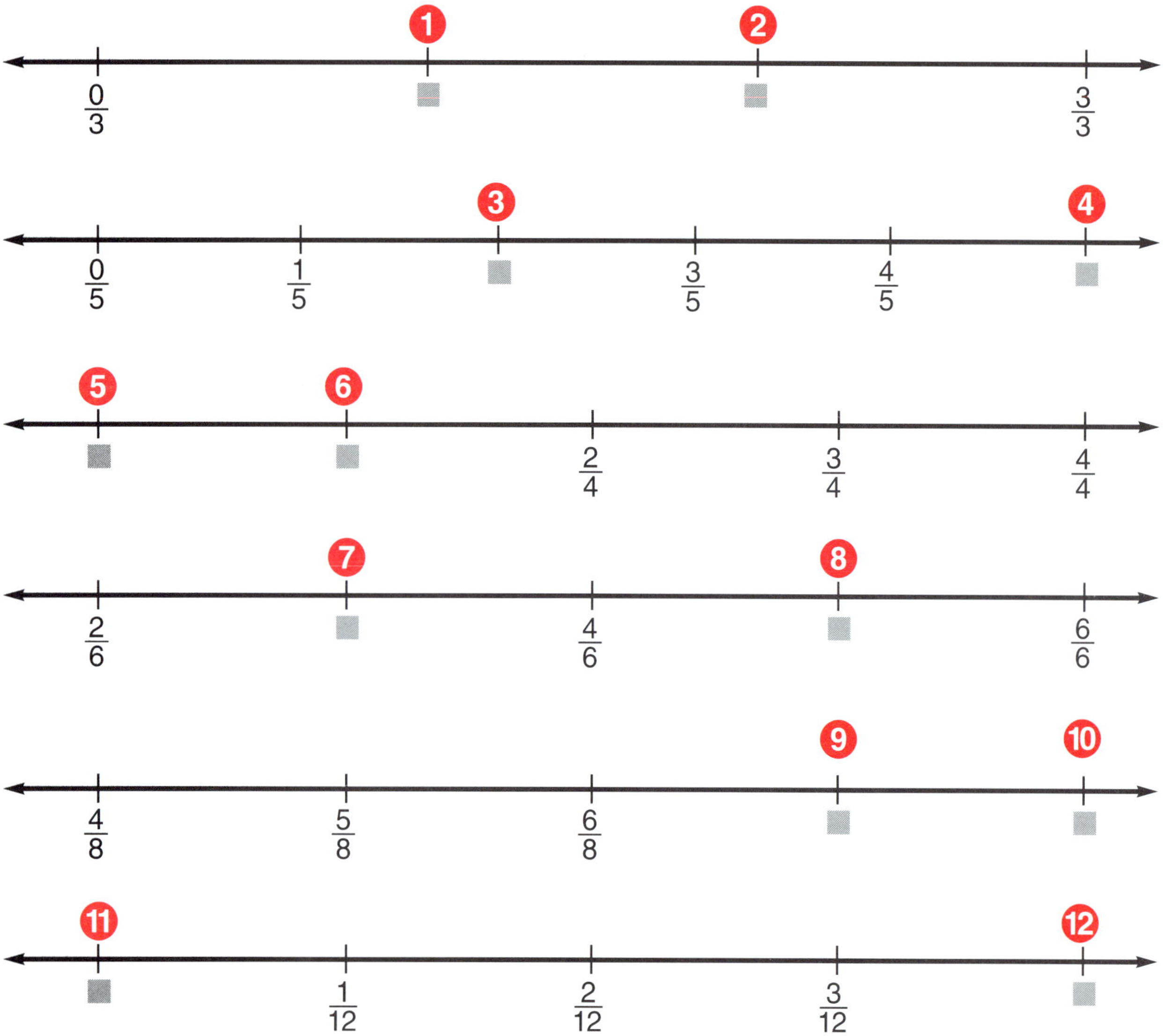

Answer Box

A	B	C	D	E	F
$\frac{0}{12}$	$\frac{0}{4}$	$\frac{5}{6}$	$\frac{4}{12}$	$\frac{3}{6}$	$\frac{8}{8}$
G	**H**	**I**	**J**	**K**	**L**
$\frac{2}{5}$	$\frac{7}{8}$	$\frac{2}{3}$	$\frac{1}{4}$	$\frac{1}{3}$	$\frac{5}{5}$

 Objective: Use a number line to order fractions.

Mix-Up

$1\frac{3}{4}$ is a **mixed number**.

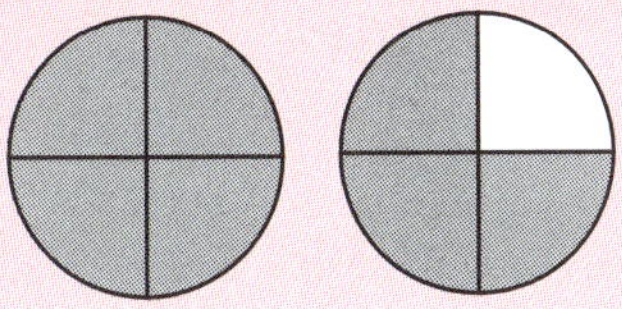

Find the mixed number that tells how much is shaded.

1 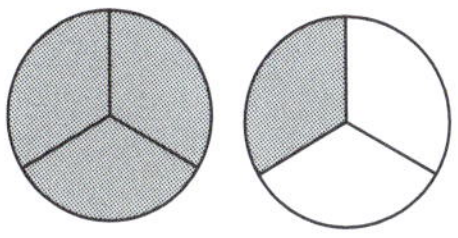

2 

3 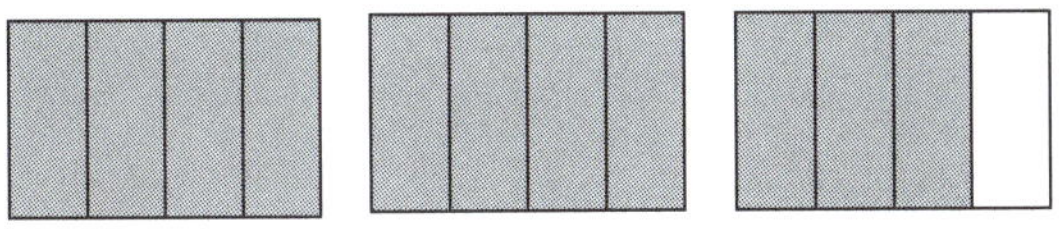

4 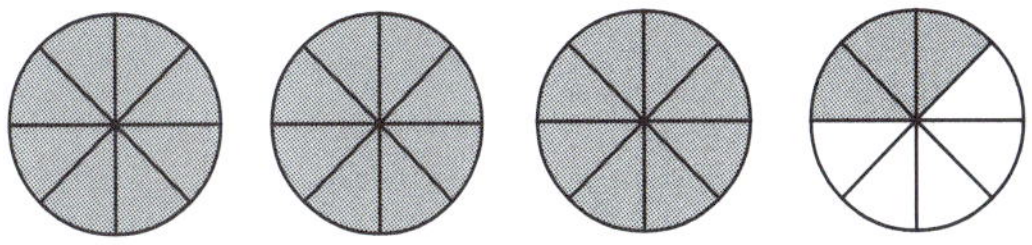

5 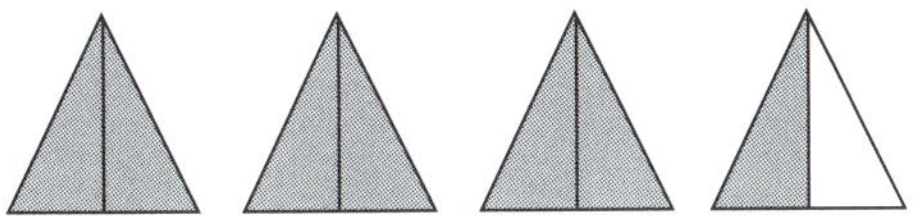

6 

7 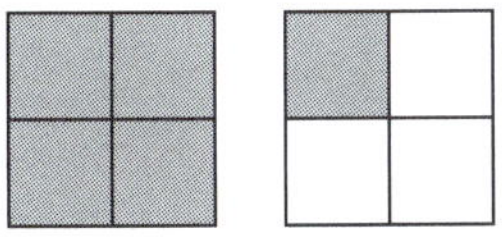

8 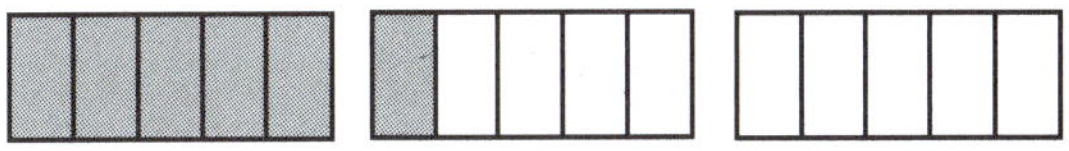

9 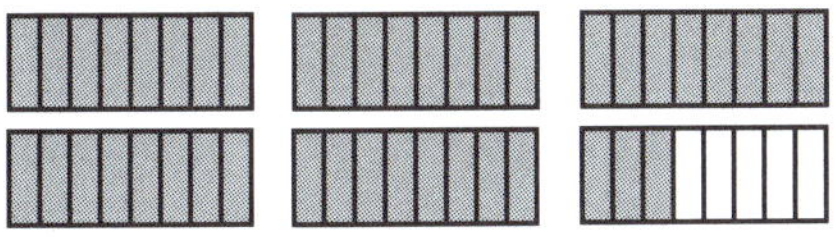

10 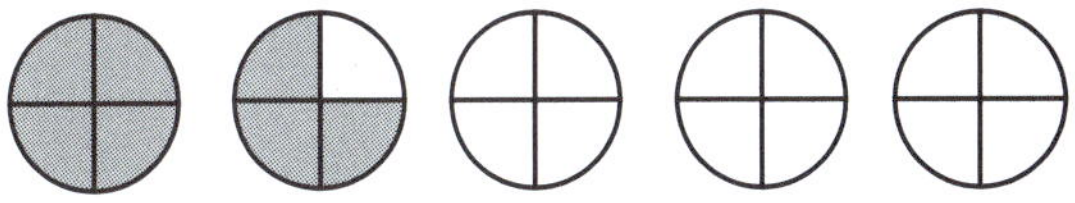

11 

12 

Answer Box

A	B	C	D	E	F
$4\frac{1}{4}$	$1\frac{1}{5}$	$5\frac{3}{8}$	$2\frac{3}{4}$	$1\frac{3}{4}$	$1\frac{1}{4}$
G	**H**	**I**	**J**	**K**	**L**
$1\frac{4}{5}$	$2\frac{2}{3}$	$3\frac{1}{2}$	$2\frac{1}{3}$	$3\frac{3}{8}$	$1\frac{1}{3}$

Objective: Name a mixed number, given a regional model.

Mix and Match!

Example

Add to find the mixed number.

$2 + \frac{7}{8} = 2\frac{7}{8}$

Add to find the mixed number.

1. $4 + \frac{1}{2}$
2. $1 + \frac{3}{4}$
3. $4 + \frac{2}{3}$
4. $3 + \frac{5}{6}$
5. $5 + \frac{3}{8}$
6. $2 + \frac{1}{3}$
7. $4 + \frac{5}{6}$
8. $1 + \frac{4}{5}$
9. $2 + \frac{3}{4}$
10. $3 + \frac{8}{9}$
11. $3 + \frac{5}{8}$
12. $3 + \frac{1}{2}$

Answer Box

A	B	C	D	E	F
$1\frac{3}{4}$	$3\frac{1}{2}$	$4\frac{1}{2}$	$3\frac{5}{6}$	$2\frac{3}{4}$	$5\frac{3}{8}$
G	**H**	**I**	**J**	**K**	**L**
$1\frac{4}{5}$	$2\frac{1}{3}$	$4\frac{5}{6}$	$3\frac{8}{9}$	$3\frac{5}{8}$	$4\frac{2}{3}$

 Objective: Name a mixed number given a whole number plus a fraction.

Made in the Shade

Example

Find the missing number.

___ tenths is shaded.
2 tenths is shaded.

___ is shaded.
0.2 is shaded.

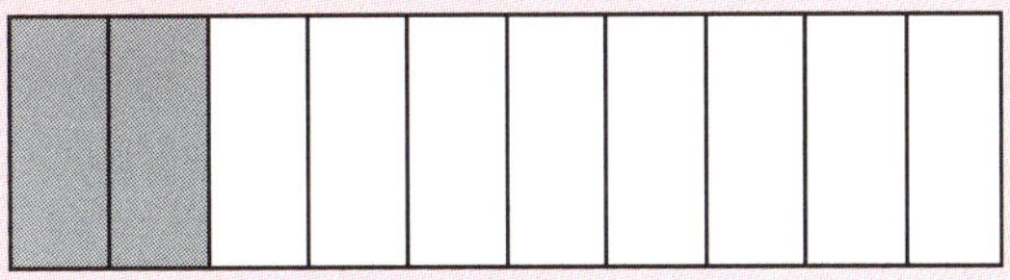

Find the missing number.

1. 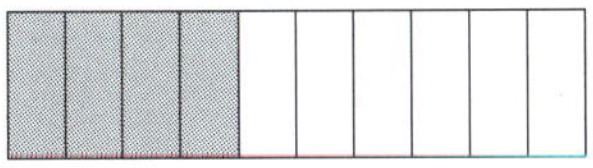
   ___ tenths is shaded.

2. 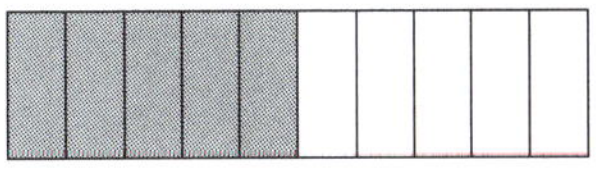
   ___ tenths is shaded.

3. 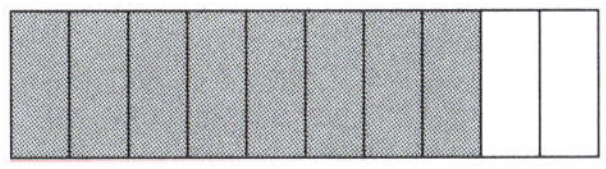
   ___ tenths is shaded.

4. 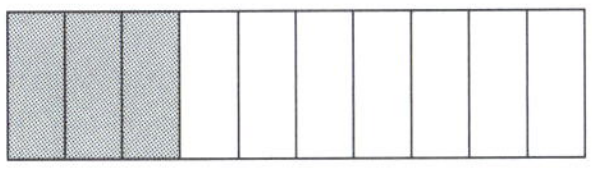
   ___ tenths is shaded.

5. 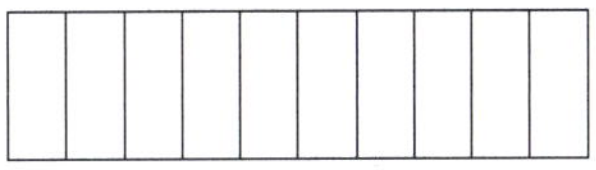
   ___ tenths is shaded.

6. 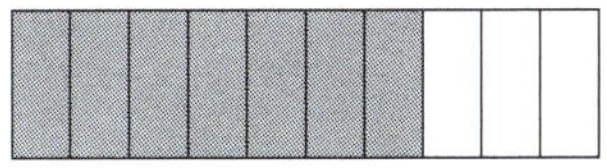
   ___ tenths is shaded.

7. 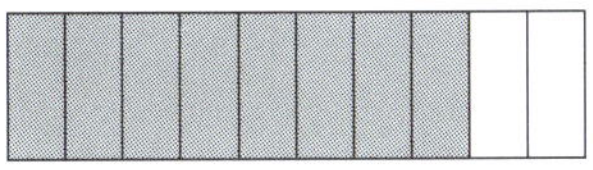
   ___ is shaded.

8. 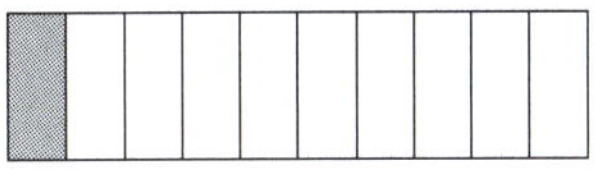
   ___ is shaded.

9. 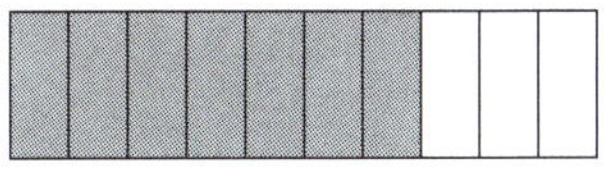
   ___ is shaded.

10. 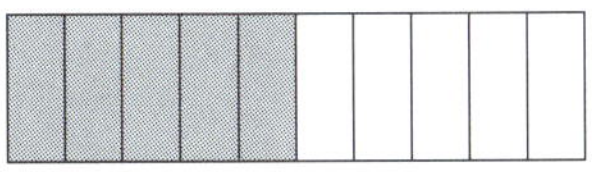
    ___ is shaded.

11. 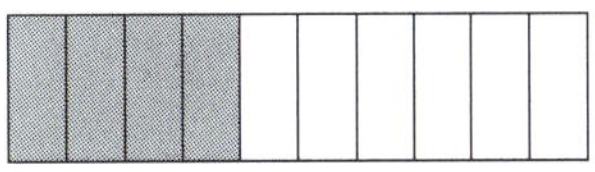
    ___ is shaded.

12. 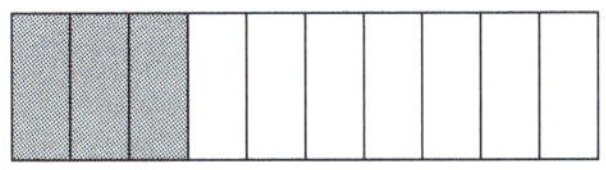
    ___ is shaded.

Answer Box

A	B	C	D	E	F
3	4	0.1	0.3	0.7	8
G	**H**	**I**	**J**	**K**	**L**
0.5	0.8	5	7	0	0.4

Square Shades

Example

Find the decimal that describes how much is shaded.

Think: 42 out of 100 squares are shaded.

So, **0.42** is shaded.

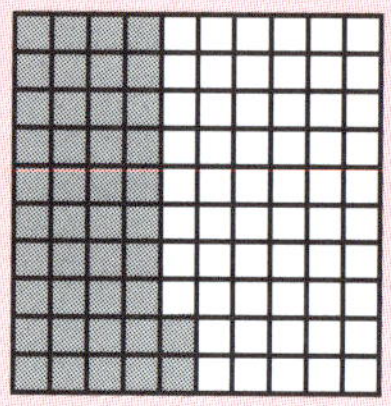

Find the decimal that describes how much is shaded.

1. 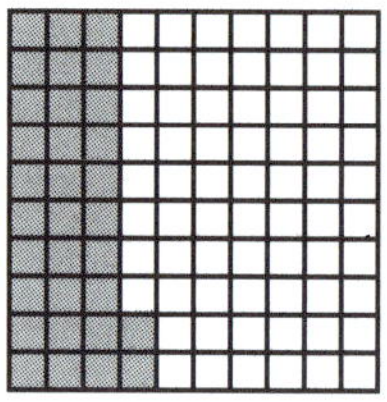

2. 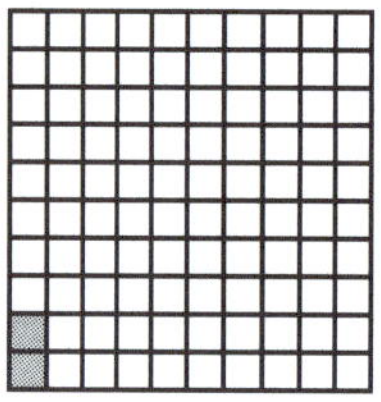

3. 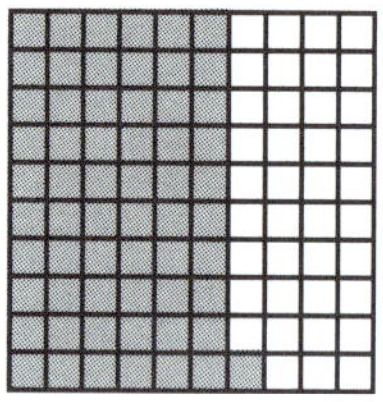

4. 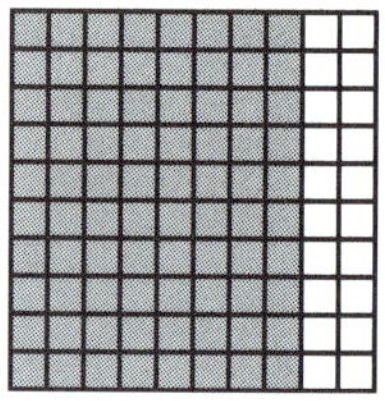

5. 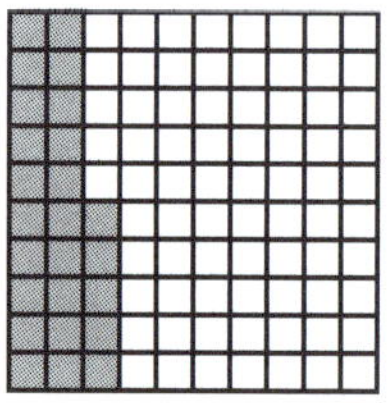

6. 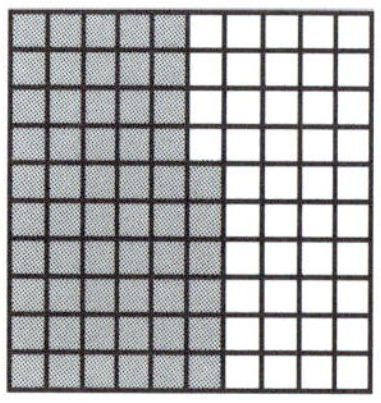

7. 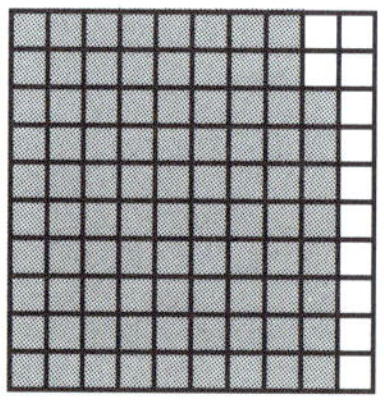

8. 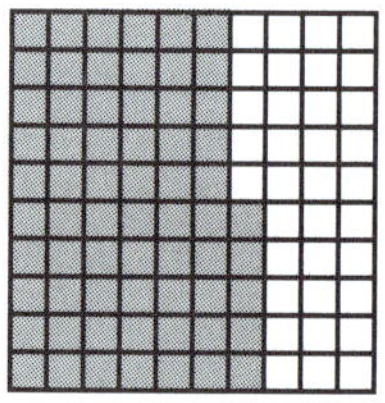

9. 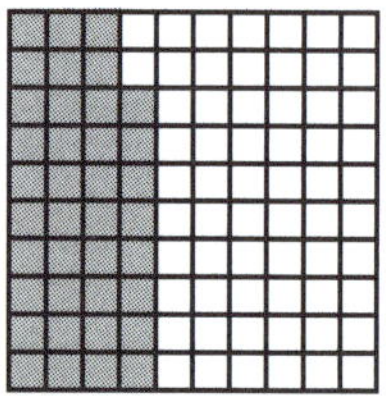

10. 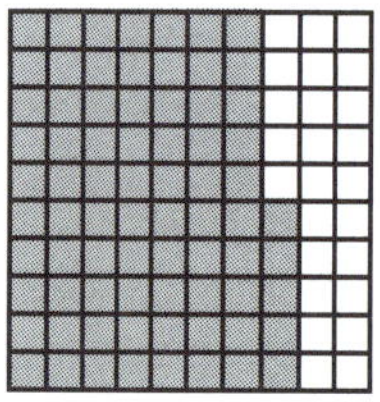

11. 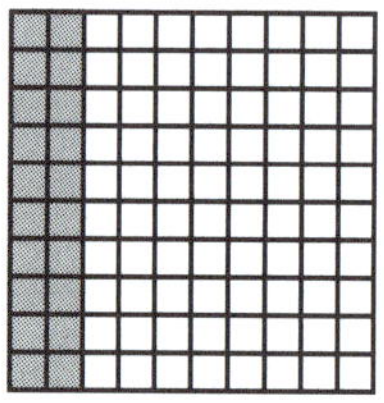

12. 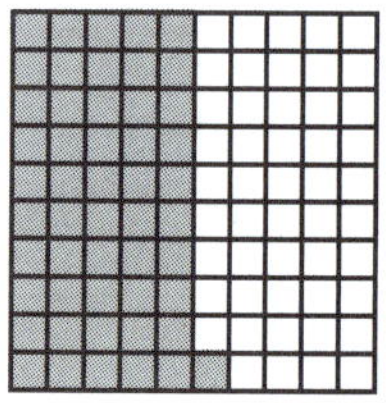

Answer Box

A	B	C	D	E	F
0.38	0.61	0.80	0.32	0.56	0.02
G	**H**	**I**	**J**	**K**	**L**
0.25	0.20	0.88	0.75	0.65	0.51

Objective: Name a decimal in hundredths.

What's My Decimal Name?

$\frac{4}{10}$ = 4 tenths = 0.4

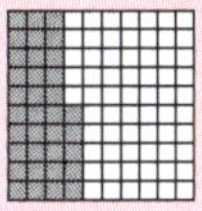

$\frac{35}{100}$ = 35 hundredths = 0.35

Find the decimal that is equal to the fraction.

1. $\frac{7}{10}$
2. $\frac{3}{100}$
3. $\frac{58}{100}$
4. $\frac{45}{100}$
5. $\frac{9}{100}$
6. $\frac{4}{10}$
7. $\frac{85}{100}$
8. $\frac{1}{100}$
9. $\frac{3}{10}$
10. $\frac{1}{10}$
11. $\frac{7}{100}$
12. $\frac{99}{100}$

Answer Box

A	B	C	D	E	F
0.1	0.85	0.3	0.58	0.01	0.99
G	**H**	**I**	**J**	**K**	**L**
0.45	0.7	0.09	0.07	0.4	0.03

Objective: Name a decimal less than 1 for a given fraction.

My Name Is Greater Than 1!

Example

Find the decimal that is equal to the mixed number.

$5\frac{2}{10} = \mathbf{5.2}$ $15\frac{47}{100} = \mathbf{15.47}$

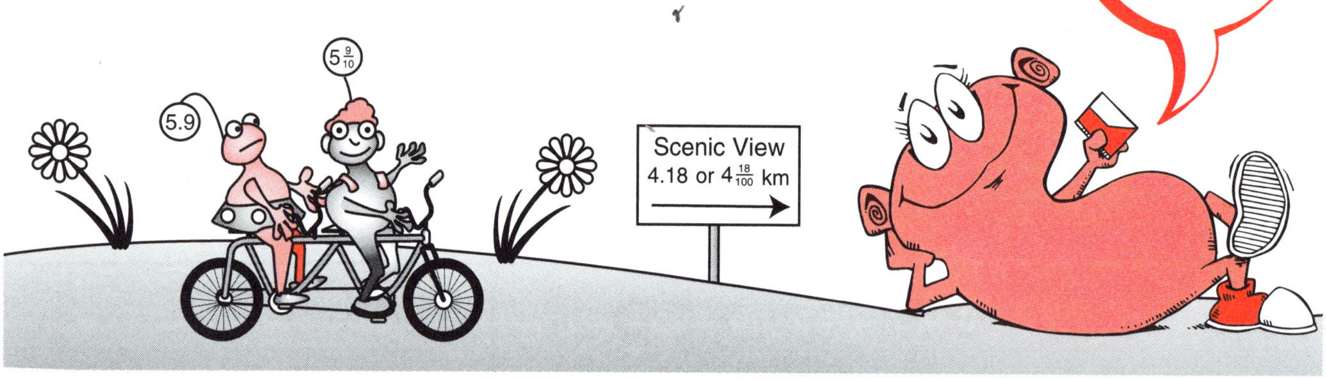

Find the decimal that is equal to the mixed number.

1. $2\frac{3}{10}$
2. $4\frac{57}{100}$
3. $7\frac{8}{10}$
4. $2\frac{33}{100}$
5. $51\frac{8}{10}$
6. $20\frac{3}{10}$
7. $45\frac{7}{10}$
8. $50\frac{18}{100}$
9. $7\frac{8}{100}$
10. $5\frac{18}{100}$
11. $45\frac{7}{100}$
12. $70\frac{8}{10}$

Answer Box

A	B	C	D	E	F
4.57	20.3	45.07	51.8	2.3	2.33
G	**H**	**I**	**J**	**K**	**L**
50.18	70.8	7.8	7.08	5.18	45.7

Objective: Name a decimal greater than 1 for a given mixed number.

Where to Compare?

Choose the decimal that makes the statement true.

1. 2.5 < ■ — 2.3 or 2.7
2. 0.6 > ■ — 0.4 or 0.7
3. 2.4 > ■ — 0.7 or 2.7
4. 4.83 > ■ — 6.83 or 2.83
5. 6.83 = ■ — 6.63 or 6.83
6. 2.87 < ■ — 2.83 or 2.97
7. 2.9 > ■ — 2.3 or 3.1
8. 9.26 > ■ — 0.66 or 9.72
9. 4.6 < ■ — 0.4 or 6.1
10. 9.27 < ■ — 9.72 or 2.97
11. 4.1 > ■ — 6.1 or 3.1
12. 6.36 < ■ — 6.63 or 0.66

Answer Box

A	B	C	D	E	F
2.3	2.7	0.7	6.1	2.97	0.66
G	**H**	**I**	**J**	**K**	**L**
2.83	9.72	3.1	6.83	6.63	0.4

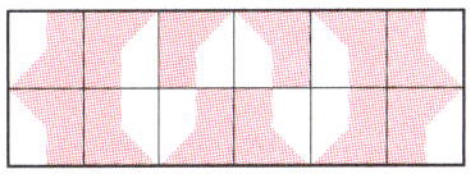

Choose the decimal that makes the statement true.

1. 0.3 < ■ — 0.5 or 0.1
2. 3.2 > ■ — 3.1 or 3.5
3. 0.89 = ■ — 0.98 or 0.89
4. 1.02 > ■ — 1.22 or 1.01
5. 5.8 < ■ — 7.2 or 0.5
6. 0.89 > ■ — 0.89 or 0.75
7. 3.22 > ■ — 3.48 or 1.22
8. 0.7 > ■ — 0.1 or 0.7
9. 0.8 > ■ — 0.7 or 7.2
10. 2.33 < ■ — 1.01 or 3.44
11. 3.48 = ■ — 3.48 or 3.44
12. 3.3 < ■ — 3.1 or 3.5

Answer Box

A	B	C	D	E	F
3.5	0.1	3.44	1.22	7.2	3.48
G	**H**	**I**	**J**	**K**	**L**
3.1	0.75	0.5	1.01	0.7	0.89

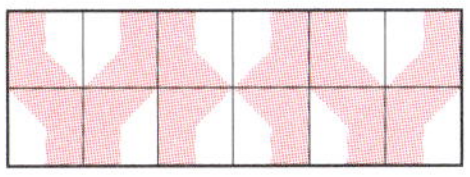

Keep Your Decimals in Line!

Find the missing decimal.

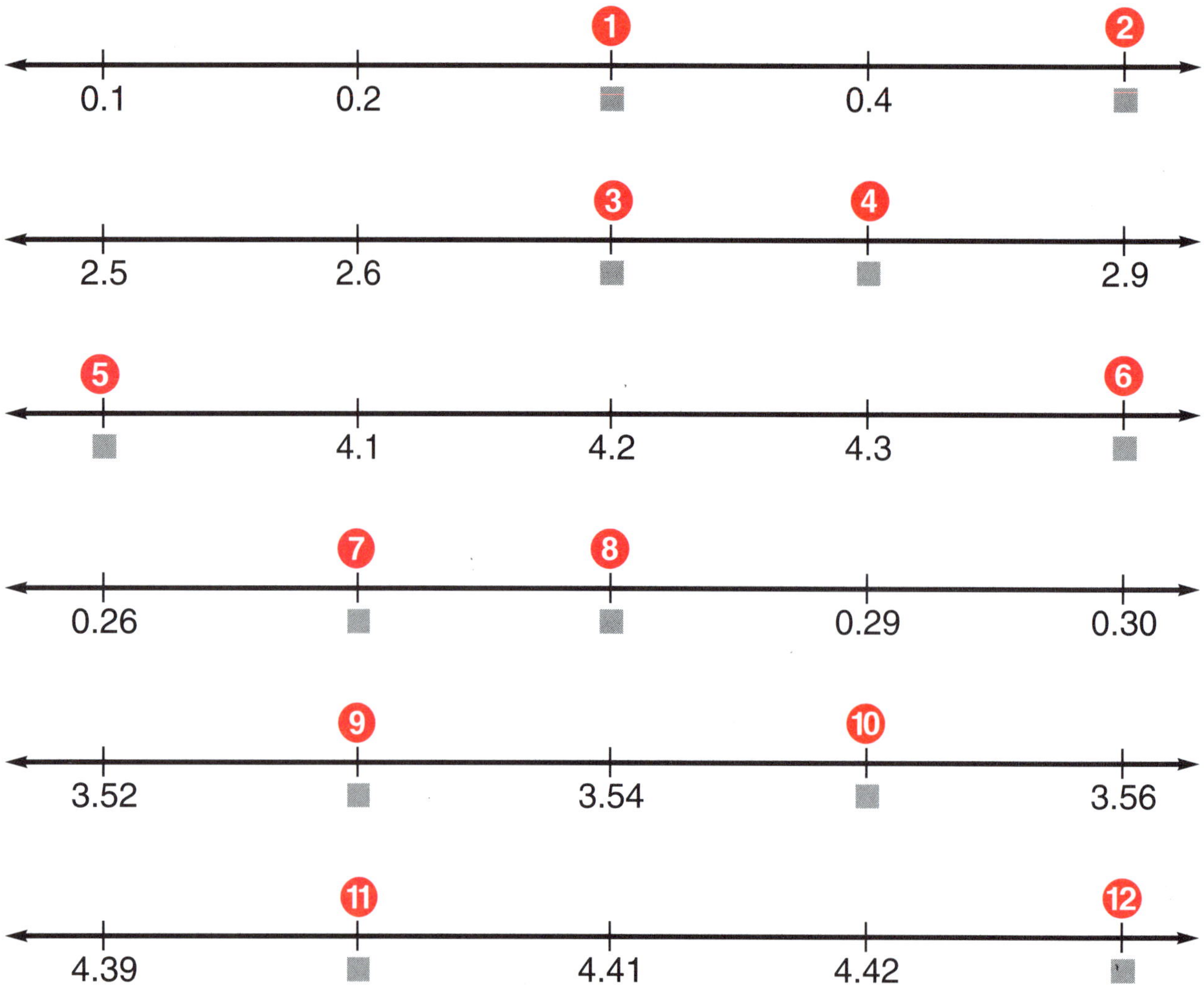

Answer Box

A	B	C	D	E	F
2.8	0.3	0.28	4.4	2.7	3.53
G	**H**	**I**	**J**	**K**	**L**
0.27	3.55	0.5	4.43	4.0	4.40

Objective: Use a number line to order decimals.